NATURAL DISASTERS, WHAT & WHY?:

1ST GRADE GEOGRAPHY SERIES

Speedy Publishing LLC
40 E. Main St. #1156
Newark, DE 19711
www.speedypublishing.com

A natural disaster is a major adverse event resulting from natural processes of the Earth.

A tropical cyclone is a circular storm that forms over warm oceans. When it hits land, it brings heavy rains and strong winds.

Hurricanes form when warm moist air over the water rises, it is replaced by cooler air. The cooler air will then warm and start to rise. This cycle causes huge storm clouds to form.

An earthquake is the shaking of the surface of the Earth, which can be violent enough to destroy major buildings and kill thousands of people.

Earthquakes happen when two large pieces of the Earth's crust suddenly slip. This causes shock waves to shake the surface of the Earth in the form of an earthquake.

A tornado consist of a very fast rotating column of air that usually forms a funnel shape.

Tornadoes tend to form from very tall thunderstorm clouds called cumulonimbus clouds.

A volcanic eruption is the point in which a volcano is active and releases lava and poisonous gasses in to the air.

Volcanoes erupt because of density and pressure.

Tsunamis are large and powerful ocean waves that grow in size as they reach the shore.

A tsunami is caused by the displacement of a large volume of water. Most tsunamis are caused by earthquakes.

An avalanche is a rapid flow of snow down a sloping surface.

Avalanches occur when large amounts of snow fall from steep slopes due to a fissure in the snow pack or due to an accumulation of too much snow for the slope's angle.

Drought is a period of below-average precipitation in a given region resulting in prolonged shortages in its water supply.

Drought may occur almost anywhere in the world.